Stefano Iannacone

Particle deposition in membranes feed channels

AF141107

Stefano Iannacone

Particle deposition in membranes feed channels

A comparison between micro-structured and commercial reverse osmosis membranes

LAP LAMBERT Academic Publishing

Impressum / Imprint

Bibliografische Information der Deutschen Nationalbibliothek: Die Deutsche Nationalbibliothek verzeichnet diese Publikation in der Deutschen Nationalbibliografie; detaillierte bibliografische Daten sind im Internet über http://dnb.d-nb.de abrufbar.

Alle in diesem Buch genannten Marken und Produktnamen unterliegen warenzeichen-, marken- oder patentrechtlichem Schutz bzw. sind Warenzeichen oder eingetragene Warenzeichen der jeweiligen Inhaber. Die Wiedergabe von Marken, Produktnamen, Gebrauchsnamen, Handelsnamen, Warenbezeichnungen u.s.w. in diesem Werk berechtigt auch ohne besondere Kennzeichnung nicht zu der Annahme, dass solche Namen im Sinne der Warenzeichen- und Markenschutzgesetzgebung als frei zu betrachten wären und daher von jedermann benutzt werden dürften.

Bibliographic information published by the Deutsche Nationalbibliothek: The Deutsche Nationalbibliothek lists this publication in the Deutsche Nationalbibliografie; detailed bibliographic data are available in the Internet at http://dnb.d-nb.de.

Any brand names and product names mentioned in this book are subject to trademark, brand or patent protection and are trademarks or registered trademarks of their respective holders. The use of brand names, product names, common names, trade names, product descriptions etc. even without a particular marking in this work is in no way to be construed to mean that such names may be regarded as unrestricted in respect of trademark and brand protection legislation and could thus be used by anyone.

Coverbild / Cover image: www.ingimage.com

Verlag / Publisher:
LAP LAMBERT Academic Publishing
ist ein Imprint der / is a trademark of
OmniScriptum GmbH & Co. KG
Heinrich-Böcking-Str. 6-8, 66121 Saarbrücken, Deutschland / Germany
Email: info@lap-publishing.com

Herstellung: siehe letzte Seite /
Printed at: see last page
ISBN: 978-3-659-81513-3

"Innovative work takes time to complete. But when it is far enough to stimulate others, it should be shared."

J. O'Connell

Alla mia famiglia, grazie per il Vostro amore e supporto.

FRAMEWORK

This report ends up the thesis project of the Master's Program in Water Technology (Chemical Engineering) from Wageningen University, being held in the premises of Wetsus, Centre of Excellence for Sustainable Water Technology, Leeuwarden, the Netherlands.

The principal target of this assignment was to study the dynamics of particle deposition in membrane feed channels, under cross-flow conditions. This project has been divided in two parts:

The first part has been carried out at the University of Twente, Faculty of Science and Technology within the *SFI* (Soft matters, Fluidics and Interfaces,) group, under the supervision of Prof. Rob Lammertink, and aimed to fabricate and characterize a new membrane concept with integrated microstructures in order to mimic the nodes of net shaped spacers. Photolithography had been used to produce membranes' patterns negatives.

The second part of this thesis has been performed at Delft University of Technology (TU Delft), Faculty of Applied Sciences within the *EBT* (Environmental Biotechnology) group, under the supervision of Prof. Cristian Picioreanu and the co-supervision of Prof. Hans Vrouwenvelder. The objective was to visually characterize particle deposition on membranes with integrated structures and on commercial membranes and spacers, under cross flow condition. Further on, an attempt to describe the experimental results with a three-dimensional numerical simulation was made in order to explain the main factors governing particle attachment.

Nomenclature	
d_{end}	Final thickness of the membrane
d_{cast}	Height of the blade of the casting knife
d_p	Particle diameter
dp	Pressure difference
m_p	Mass of the particle
L_x, L_y, L_z	Length of the domain in the three dimensional coordinates
N_p	Number of particles
p	Pressure
R_a	Average roughness
R_q	Average of height deviations
Re	Reynolds number
S_k	Shrinkage
St	Stokes number
T_{cast}	Height of the blade of the casting knife
$\boldsymbol{u}$ (u_x, u_y, u_z)	Velocity vector with the three spatial component
u_{in}	Average inlet velocity
Δt	Time step
ρ_p	Density of the particles
ρ_w	Density of water
∇	Gradient
ε_L	Porosity
η	Water dynamic viscosity
λ	Power of the van der Waals forces
x, y, z	Spatial coordinates
τ	Shear stress

TABLE OF CONTENTS

FRAMEWORK 2

1.   INTRODUCTION 5

2.   SCOPE 9

3.   MATERIALS AND METHODS 10

   PART I: Microstructured Membranes 10

   Materials, solutions and membrane preparation 10

   Membranes characterization 12

   PART II: Particles Deposition Set-Up 14

   Direct microscopic observation module for particles deposition 14

   Microspheres 16

   Microstructured membranes and commercial membranes 17

   PART III: Three-Dimensional Numerical Simulation 18

   Model geometries and computational domains 18

   Hydrodynamics 19

4.   RESULTS AND DISCUSSION 24

   3.1 MEMBRANE CHARACTERIZATION 24

   The effect of non-Solvent 24

   The Effect of PVP 28

   3.2 PARTICLE DEPOSITION 32

   Particle attachment in time 32

   The effect of feed spacer geometry 34

   Commercial net-shaped spacers 35

   Microstructured spacers 37

   The effect of the liquid velocity 39

   Correlation between flow field and particle deposition 40

5.   CONCLUSIONS AND RECOMMENDATIONS 42

BIBLIOGRAPHY 44

ACKNOWLEDGMENTS 48

1. INTRODUCTION

The availability of good quality water is vital to mankind. Population growth and economic development are driving significant increases in water demand. This trend has led to a rising interest in alternative sources for water production, drinking water *in primis* (1). One of the main sources comes from the desalination of salty water, which represents approximately 98% of the water available on the Planet.

Membrane-based technologies are among the most advanced methods in water treatment and desalination. Reverse osmosis (RO) is the key process to desalinate seawater and brackish water and it has been used for more than 30 years (2), while nanofiltration (NF) membranes are used to remove organic matter and hardness from surface and ground water. For these two different types of water treatments, spiral wound membrane modules are primarily employed (1).

Particularities of these modules are the presence of spacers in the feed and permeate channels. The feed flow spacers, which usually have the form of woven crossed cylinders, serve mainly to separate adjacent membrane sheets and create flow passages, but also to promote mixing and thus enhance mass transport (3; 4). In recent years the important role of membrane spacers has been recognized and many experimental and theoretical studies were accomplished, aiming at understanding the underlying phenomena and designing more efficient spacer geometries (3; 5). Clearly, an improvement in spacer configuration has to be evaluated on the basis of three major aspects. The first is the enhancement of mass transport, which reduces

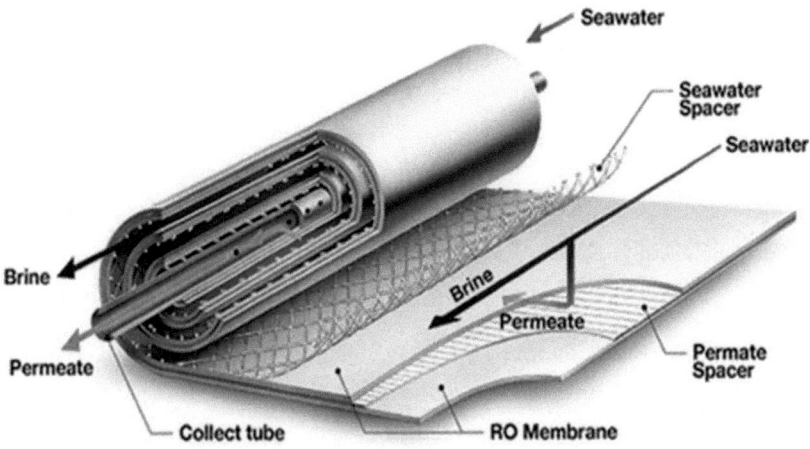

Figure 1: Spiral wound membrane element configuration (56)

concentration polarization and osmotic pressure at the membrane surface, as well as the potential for scaling (5). The second is the enhancement of shear stresses at the membrane surface in order to minimize the tendency of fouling species to deposit and reduce membrane flux and rejection or modify the feed channel hydrodynamics (6). The third aspect is the pressure drop, associated with the energy required for pumping and contributing to the decreasing membrane productivity along an array of elements (3; 7; 8). The latter is more prominent in low pressure membrane applications (i.e. nanofiltration, UF). In general, however, enhanced mass transfer and shear comes at the expense of higher energy input and higher pressure drop. Therefore, a unique answer to such conflicting requirements does not exist and tailor-made spacers could be developed for different types of membranes or applications and feed water types.

Recently, the Lammertink group from Twente University - the Netherlands - presented a new versatile method to fabricate polymeric flat sheet membranes with integrated "free standing" structures, acting as feed spacers, on a microstructured silicon mold (9; 10; 11). The geometry of these structures can be easily tuned depending on the fouling propensity of the feed solution and the membranes application (microstructured hollow fibers membranes and microsieves can be fabricated as well) (12; 13). The positioning and orientation of the structures can be used to enhance surface shear within the spacer filled channel. Not less important, there also may be the added advantage of the permeability of these structures, providing enhanced surface area available for filtration (9; 13).

Four types of fouling can be distinguished and generally they do not occur independently (6; 14): mineral scaling, organic fouling, colloidal fouling and biofouling. Deleterious consequences of all types of fouling include increased hydraulic resistance, decreased membrane permeability, concentration polarization, less salt rejection besides the increased operational costs (7; 6; 8; 9; 15). Despite all the massive progresses in chemical products, design and materials, the undesirable (bio)fouling phenomenon is only diminished, but still not completely controllable. Fouling and biofouling are indeed considered the Achille's heel of all membrane processes (7; 16).

It is well known in this respect that the pattern of fouling deposits is irregular. Several reports suggested that the net biomass accumulation pattern closely follows the geometry of spacers (8; 15; 17; 18). It is recognized that initial microbial cell deposition is a critical stage in the overall process of biofouling. It has been reported that primary adhesion is governed mainly by physical-chemical factors such as hydrodynamics, solution chemistry and interfacial forces

(16). Bacteria are generally associated with an interface, they are colloidal in nature and they have many similarities with their inorganic counterparts with respect to membranes fouling. Therefore, in the first fouling stages microorganisms may be regarded as bio-colloids (19; 20).

An overview of the current status of fouling research is given in the following two paragraphs.

Experimental methods

There have been an increasing number of studies on RO and NF colloidal fouling in the last decades, focusing mostly on fouling indexes. For example, the forces between charged particles interacting within a liquid medium and their adhesion to surfaces have been theoretically described (21; 22; 23). Laboratory experiments have been carried out under controlled hydrodynamic and water chemistry conditions using model colloidal/macromolecular foulants for RO systems (14; 24; 25). A comprehensive review of the colloidal interactions and fouling of NF and RO was given by Tang et al. (14). They suggested that the formation of a colloidal cake layer can severely reduce the membrane flux as a result of the cake layer hydraulic resistance and/or the cake-enhanced osmotic pressure (CEOP). Furthermore, fouling is strongly affected by the feed water composition, the hydrodynamic conditions and membrane properties. However, most of the studies did not include the effect of feed spacer on fouling. One of the simplest approaches to study fouling on membrane systems with spacers is the non-invasive direct microscopic observation (18; 26).

The study of fouling of microstructured membranes was initiated by Ngene et al. (9). They showed a tendency for all types of fouling to develop upstream the microstructured pillars rather than on the membrane. This trend was observed independently of both the structure geometry and orientation and the liquid velocity and local wall shear, conflicting with previous reports on commercial feed spacers (7; 27).

Computational methods

Computational fluid dynamics (CFD) has been used for long time now as a tool to model fluid flow through the feed channel in membrane separation processes. There is abundant literature in this field for 2d and 3d channel geometries, with or without feed spacers (3; 4; 5). Recently, numerical models for fouling and its effect on process performance have been proposed, considering the influence of feed spacer geometries, bacteria detachment and the effect of filtration (6; 8). Li et al. presented studies of flow and mass transfer by performing 3d direct numerical simulations in a geometry representing the curvature of spiral wound module with spacers. They showed that particle deposition on the membranes in the empty channel was

7

dependent on the shear stress (15; 28). A comprehensive review of modeling particulate movement in fluids has been presented by Marshall et al. (29; 30). He proposed a multiple-time step computational approach for efficient discrete-element (using Lagrangian particle methods (31)) modeling of aerosol flows containing adhesive solid particles. Figure 2 presents the relation between characteristic time and length scales for various particle simulation methods, along with typical applications. Undoubtedly, the particles in these systems represent a wide range of discrete physical elements, such as atoms, molecules, nuclei, biological cells, aerosol or colloidal particles and granules. The methods proposed by Marshall include all the possible interaction forces that act on these entities. It is important to choose wisely the physics included in the model and the level of detail needed, avoiding uselessly high computational demands. Particle trajectory and behavior near planar surfaces has also been investigated with mathematical methods aiming to predict the deposition of colloidal systems in the presence of shear flow (32). It is clear that the understanding of the interactions between fluid flow, particle trajectories, spacer geometry and surfaces is of great importance. Models considering these interactions can be useful tools to predict the initial bacteria adhesion in different systems such as membrane modules and porous media. As a substitute for bacteria, abiotic particles with diameters ranging from 1 to 5 μm can be used to mimic biological cells at their first stage of adhesion (therefore, not considering their multiplication). Computational approaches can complement empirical studies and potentially improve the design of the system, if one could evaluate by numerical means the importance and relative contribution of different forces acting on particles motion and deposition.

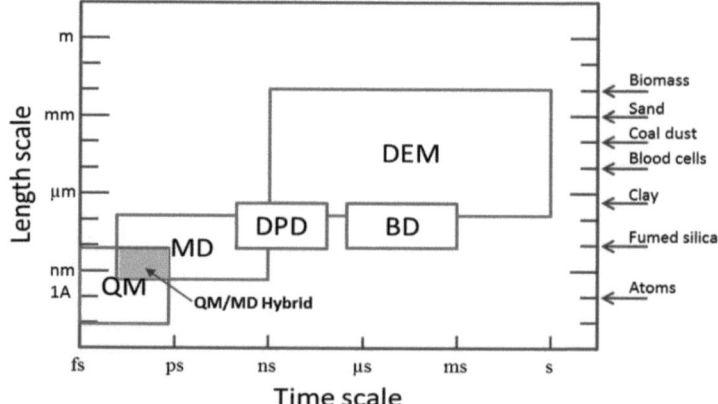

Figure 2: Diagram of time and length scales for particle flow simulation methods, including quantum mechanics (QM), molecular dynamics (MD), dissipative particle dynamics (DPD), Brownian dynamics (BD), and discrete-element method (DEM) (30).

2. SCOPE

The first part presents the preparation method and the characterization of membranes with tailored integrated shapes acting as spacing obstacles. In the second part, we studied the evolution of particle deposition as a function of time for the different spacer geometries and orientation via a non-invasive direct microscopic observation. Finally, a three-dimensional model describing the liquid flow together with deposition of particles in the feed channel of those membranes has been developed.

3. MATERIALS AND METHODS

PART I: Microstructured Membranes

Phase separation micromolding (PSμM) is a process in which a homogeneous polymer solution is cast onto a structured template (mold) (10). PSμM is very versatile because basically all the polymers can be patterned with this technique; the only matter is to find a suitable solvent and non-solvent (13). The mold used in this study is a "negative" structure, which means that the features are deeper than the starting layer (Figure 3). Each structure was etched to a depth of 250 μm. When casting a polymer solution, the height of the "blade" of the casting knife is usually measured starting from the unstructured layer of the molds. The polymer solutions are cast and fill the features in the mold. Upon solidification, the features on the support are replicated and the membrane can be removed from the mold. Using this method, the desired structures can be well replicated and the properties of the membrane can be easily tuned (12). The process is briefly schematized in Figure 3. Two different polymeric structures (the four-tipped star and the teardrop) obtained with this method will be presented in this thesis.

Materials, solutions and membrane preparation

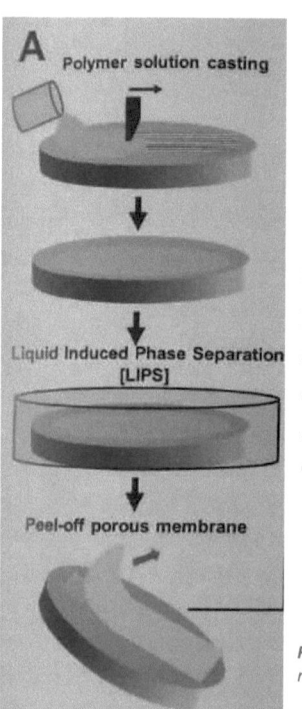

The polymer used in these experiments was polyethersulfone (PES, Ultrason, E6020P). The additive for the polymer solutions was polyvinylpyrrolidone (PVP K12 and K30, Fluka). N-methylpyrrolidone (NMP, 99%, Acros Organics) was used as solvent. Tap water and a mixture of tap water and NMP were used as non-solvent. All reagents were used as received, without further purification. Unless otherwise specified, solutions were prepared by weighing all the components into a plastic bottle and left on a rolling bank until complete dissolution. Solutions composition, solvent, heights of the casting knife and coagulation baths used for the experiments are summarized in Table 1.

Figure 3: Illustration of fabrication method for microstructured membranes by PSμM

Table 1: Composition/characteristics of the solutions and casting/coagulation conditions

Polymer	Additive	Solvent	Composition P/A/S * (w/w %)	Heights of the casting knife (μm)	Coagulation Bath
PES	-	NMP	(15/0/85)	250 – 300 – 350 – 400 – 450 – 500 – 550 – 600 – 650 – 700.	Tap water and Water/NMP (50% v/v)
PES	PVP (K12)	NMP	(15/5/80)	550	Tap water and Water/NMP (50% v/v)
PES	PVP (K30)	NMP	(15/5/80)	550	Tap water and Water/NMP (50% v/v)

* P = polymer; A = additive; S = solvent.

Figure 4 (A, B, C) illustrates the main steps in preparation of a PES membrane obtained with the method described above for the teardrop shape. Microstructured membranes were prepared by spreading a certain amount of polymer solutions on silicon wafers (Figure 4–A) using a custom-made doctor blade with micrometric screws and analogue display to regulate the casting thickness (Figure 4–B). The silicon wafer was microstructured through standard photolithographic methods combined with deep reactive ion etching (DRIE) in clean-room facilities. These structures were etched to a depth of 250 μm. The mold with the cast solution on top was afterwards placed into a glass desiccator, equipped with a vacuum pump for 7 minutes (5minutes as first round then other 2 minutes) in order to take the air out from the polymer solution and accelerate the polymer deposition into the stencils. Subsequently the mold was immersed into a non-solvent bath (tap water or water/NMP 50% mixture) for coagulation and liquid induced phase separation (LIPS). The solutions became turbid and the polymer precipitated instantaneously. In some cases the membrane lifted off after some time in this bath by itself, indicating completion of the phase separation. When this was not the case, after 30 min, when the white color was uniform over the whole surface, the membrane was carefully removed from the support and rinsed with tap water overnight. All the microstructured membrane obtained were dried and stored between soft tissues in order to prevent folding and damaging of the micro-pillars.

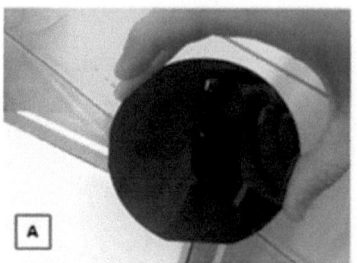
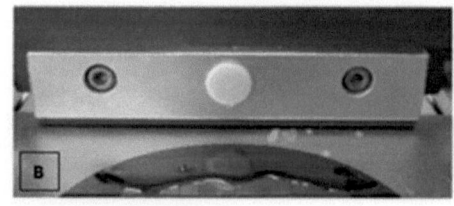
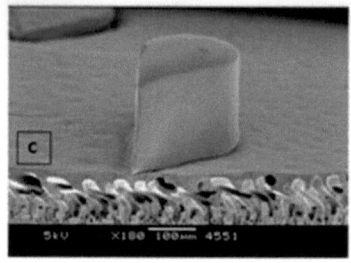

Figure 4: Steps in the preparation of membranes via PSµM. The solution is cast onto a silica mold (A and B) and after phase separation a structured membrane is obtained as seen in SEM micrograph (C) (here, the teardrop).

Membranes characterization

For surface characterization, membranes were cast at 550 microns' height, both for PES solution without additives and for solutions containing 5% PVP (see Table 1)

AFM analyses were performed for measuring the average roughness, R_a (Eq. 1) which is the arithmetic average of the absolute values of the root mean square average of height deviations R_q (Eq. 2), taken from the mean image data plane Z.

$$R_a = \frac{1}{N}\sum_{j=1}^{N}\left|Z_i - R_q\right| \qquad \text{[Eq. 1]}$$

$$R_q = \sqrt{\frac{\sum z_i^2}{N}} \qquad \text{[Eq. 2]}$$

As representative samples, round pieces of membrane (diameter of 0.5 cm), were taken for the analysis. First the membranes were analyzed with the AFM then with SEM. The advantage to use firstly AFM is because this is not a disruptive method so the same samples can be used for further analysis. AFM images were obtained with a Nanoscope microscope (Bruker type) in tapping mode. Images were obtained at room temperature, tapping tips obtained from etched silicon having a high aspect ratio, thus avoiding or minimizing convolution of the tip shape by the membrane surface.

Due to the presence of the microstructures, which did not allow the tip to get close enough to the surface and therefore create very disturbed signal, the membranes had to be cut in the proximity of a pillar and a "side approach" scan was performed. In any case, eye inspection facilitated the process of selection of several reasonable starting point candidates whose outcomes are conveniently averaged. The scanned area for each image was chosen to have enough available surface for measurements along with a high number of pores to assure statistical relevance (33). The obtained images have been conveniently analyzed by means of the Nanoscope Analysis v1.4 (Bruker). Each photograph has a resolution of 552 × 612 pixels.

SEM pictures have been taken at room temperature with a magnification from 2000× to 2500× with the SEM, JEOL 5600 at 5 kV and analyzed with the software SEMafore. Each photograph was taken with a high resolution.

For cross-sectional analysis, membranes were cast at different heights and using two different coagulation baths (see Table 1). The morphology of the structures was observed using the same SEM apparatus described above. Pictures have been taken with a range of magnification between ×50 and ×350 and analyzed with the same software, at room temperature. To obtain a perfect breaking point, membranes were immersed in ethanol for a few seconds and then fractured in liquid nitrogen. Before entering the microscope, samples were sputtered with a thin layer of gold to improve conductivity.

PART II: Particles Deposition Set-Up

For the set of experiments described in this thesis, microstructured membranes were prepared in Twente University, Enschede and delivered to Delft University of Technology for the experiments. Membranes with a casting thickness set at 350 µm fabricated from PES+PVP(K30) with tap water coagulation bath were employed. The chemistry and the thickness of the membranes have been chosen after screening the results obtained experimentally and literature reviews (9). As for the shapes, the four-tipped star and the teardrop have been selected. The commercial membranes, together with feed and product spacers were taken from a new spiral wound RO element (DOW Filmtec BWROLE-440, 8 inch diameter).

Direct microscopic observation module for particles deposition

The scheme of the flow system and the direct observation module are illustrated in Figure 5 and 6.

The head tank, containing 400 mL of Milli-Q water, is located on an upper shelf so that the aqueous suspension of particles can reach the membrane fouling simulator (MFS) by gravity.

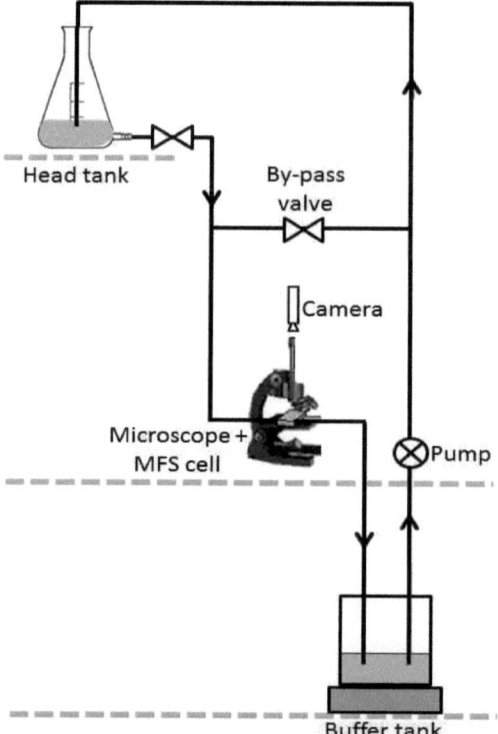

Figure 5: Installation scheme consisting in a head tank, delivering gravitationally the solution with particles to a buffer tank. During its path the solution passes through the flow cell installed under a microscope equipped with a camera. The solution from the buffer tank is recirculated into the head tank with a gear pump. The by-pass valve is used only after the installation of new membranes, in order to eliminate the micro bubbles attached to the spacer, flushing the system with clean water at a high flow rate.

The suspension is collected into a buffer tank (c.a. 200 mL). This tank was placed below the working table in which are installed the pump, the microscope and the flow cell (MFS) (see Figure 5). A suspension of red, polystyrene micro-spheres was injected into Demi-water in the head tank to reach a final particle concentration of 47 mg/L. To study the particle deposition in the MFS, this suspension was recirculated through the system for 24 hours or more, at room temperature. To bring up the liquid from the buffer tank to the head tank a digital gear pump (model Cole-Parmer Instrument) is used, at a flow rate set to keep the volumes of the liquid in the buffer and the head tanks constant. The flow system is designed in this way aiming to minimize flow pulsation, which can induce forces that influence particles trajectories and deposition. The experiment duration was long enough to allow clearly observable patterns of particle deposition on the microstructures, spacer, membrane and glass surfaces.

A newly designed membrane fouling simulator (STT Products, Tolbert - The Netherlands – Figure 6) allowed direct observation of particles deposition onto RO and microstructured membranes under cross flow conditions. The membrane fouling simulator (MFS) is constructed from two connected pieces (top and bottom parts), kept together by ten metal screws. The bottom part consists of 14 mm thick polyvinylchloride (PVC) and the top plate consists of 6 mm aluminum clamp and a 1.1 mm thick transparent glass plate with a thin rubber seal to avoid leakage. The height of the cross flow channel, in which the membrane is located, is 0.78 mm, the length is 37 mm and the width is 15 mm. An assembly of permeate spacer, membrane sample and feed spacer is placed on the bottom part of the MFS.

The porosity of the channel (ε_L) when filled with microstructured membranes was calculated to be 94% for the four-tipped star and 97% for the tear drop pattern. For the RO membrane with commercial spacer the porosity was ~89%.

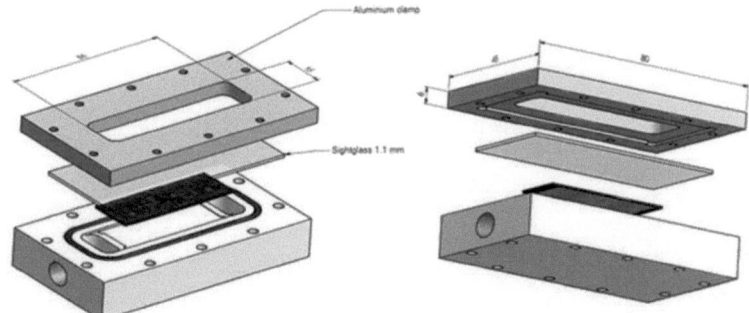

Figure 6: Scheme of the MFS components. MFS external dimensions (L/W/H): (80 x 45 x 20). Internal cross flow channel dimensions (35 x 15 x 0.78). The measures are expressed in millimeters

All the monitors were operated in cross-flow mode with no permeate production. A fixed flow rate was applied at the flow cell inlet, calculated in order to have an average cross-flow velocity of 0.14 m/s in the feed channel, for all types of membranes employed. It was assumed that the flow rate would not change significantly during the particle deposition in the MFS. The flow cell was placed on the fixed stage of a microscope (Olympus TOKYO 212684) equipped with an adjustable objective. The cell was illuminated from above with a table lamp. Pictures have been taken with a digital camera, 12.1 mega pixels of resolution (SONY Cyber-shot, 2.8-5, 8/5, 35-21.4). Images were obtained in real time during each experiment at interval of approximately 2 hours.

Microspheres

Hydrophobic, red dyed, surfactant free, spherical polymer (polystyrene) particles, with a diameter of 2 μm were used for the experiments (Carboxyl Latex, 3% w/v – Invitrogen). The microspheres are charge-stabilized by carboxylic groups, present on the particle surface (see Table 2). In the aqueous suspension used in the experiments, the electrostatic repulsion given by the carboxyl groups is sufficiently strong so that a stable suspension is obtained. The particles "feel" the repulsion over distances similar to those of attraction, so that no aggregation is observable. In addition, the carboxylic groups make the particles more hydrophilic.

Table 2: Polystyrene particles properties (34).

PROPERTY	REPORTED VALUE	CARBOXYLATION
Density	1055 Kg/m³	
Deformability	Rigid, compressive modulus 3,000 MPa	
Insoluble in	Acetone, butane, ethanol, ether, methanol, hexane, phenol, propanol, water.	
CCC* hydrophilic	>1.0 M univalent ions, pH 7	

*** *Critical Coagulation Concentration: concentration of specific ions required to produce rapid aggregation in a suspension of latex particles.*

Microstructured membranes and commercial membranes

Experiments have been carried out with both microstructured membranes and commercial membranes. In addition, the net-shaped commercial spacer was placed in different orientations in respect to the main flow direction. For commercial spacer experiments, coupons of feed, product spacer and membranes were placed in the MFS, resulting in the same channel thickness as in spiral-wound elements. For the microstructured membranes, a coupon of product spacer plus a coupon of commercial membrane were put beneath, allowing the pillars to face the top glass window and have a defined fluid pattern. Photographs of representative elements for the commercial spacer and each microstructured membrane employed are shown in Figure 7.

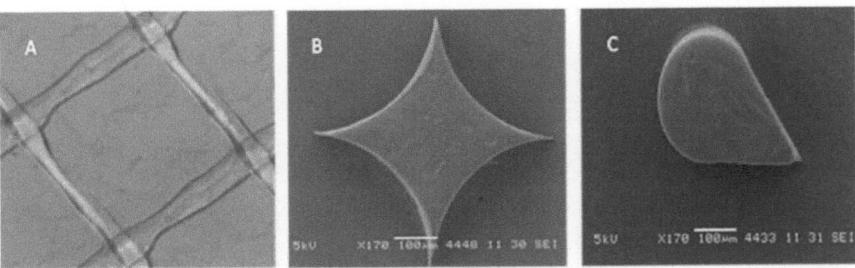

Figure 7: Geometry of obstacles present in the feed spacer channel. A) Microscopy image of a representative spacer element from commercial RO modules (DOW). SEM images of the microstructures: B) four tipped star, C) teardrop.

PART III: Three-Dimensional Numerical Simulation

We present here a three-dimensional model describing the liquid flow and particle trajectories leading to deposition of particles in a channel containing a traditional reverse osmosis feed spacer or microstructured membranes. The objective was to verify the importance of hydrodynamics on the formation of characteristic deposition patterns as obtained experimentally.

Model geometries and computational domains

The influence of channel (obstacle) geometry on particle deposition pattern was evaluated numerically. In all the simulations the computational domain contained a single representative spacer element from the whole array of identical elements. Geometries of a representative element for each microstructure and spacer employed are presented in Figure 8. These model geometries were constructed based on own measurements of the commercial feed spacer and micro-structured membranes used in the experiments. An accurate reproduction of the actual geometry (e.g., the right dimensions of the spacer filaments, height of the channel, exact angles, etc.) and correct flow velocity in the channel are crucial to obtain simulation results comparable with experimental observations (8).

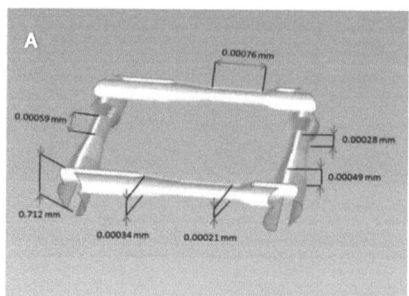
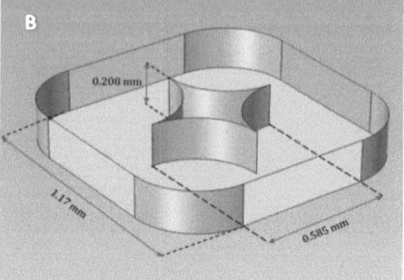
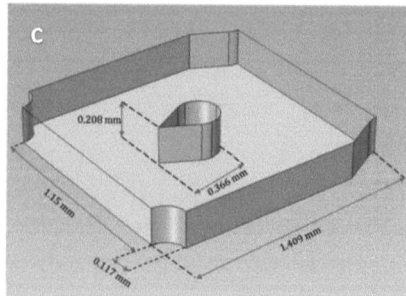

Figure 8: The modeled geometries of the DOW feed spacer (A) and the two microstructures (B and C) with geometrical dimensions.

18

Hydrodynamics

The undisturbed (particle-free) hydrodynamic field of velocities $\mathbf{u}(x,y,z) = [u_x, u_y, u_z]$ was obtained over the domain of interest by numerically solving the stationary liquid continuity and momentum equations (Navier-Stokes) for incompressible laminar flow:

$$\rho(\boldsymbol{u} \cdot \nabla)\boldsymbol{u} + \nabla p = \nabla \cdot (\eta \nabla \boldsymbol{u}) \qquad \text{[Eq. 3]}$$

$$\nabla \boldsymbol{u} = 0 \qquad \text{[Eq. 4]}$$

where $\boldsymbol{u}$ is the vector of local liquid velocity (with components u_x, u_y, u_z respectively on the three Cartesian directions x, y, z), p is the pressure, ρ is the liquid density and η is the liquid dynamic viscosity (8; 17; 32). Under typical conditions in spiral wound membrane elements, the average velocity in the channels created by two adjacent membrane leaves and separated by the spacer does not exceed 0.4 m/s and the allowable pressure drop, recommended by manufacturers, should not exceed 0.6 bar/m (35). During our experiments, the flow rate was driven by gravity at an average velocity in the flow cell inlet u_{in} = 0.14 m/s. This velocity leads to a Reynolds number (Re) of ~300, justifying thereby the assumption of stationary laminar flow.

In the numerical model, the inlet and outlet boundaries were periodically connected with $\mathbf{u}(0, y, z) = \mathbf{u}(L_x, y, z)$ and the flow was driven by an applied pressure difference between these boundaries. As the pressure drop Δp varies for different geometries, it is necessary to calculate Δp from a separate equation that matches the resulting inlet velocity with the imposed cross flow velocity u_{in}. The lateral boundaries (on the y direction) were also periodic so that $\mathbf{u}(x, 0, z)$ = $\mathbf{u}(x, L_y, z)$ and $p(x, 0, z) = p(x, L_y, z)$. The no slip boundary condition (zero velocity) was applied to all solid surfaces (membrane, spacer/pillars, glass). Periodicity was used in order to minimize wall effects and to simulate the flow pattern in a spacer element situated somewhere in the middle of the flow channel (i.e., far from the walls and from inlet/outlet, where the flow is well established).

Particles trajectories and attachment

The mathematical model for particle tracking (PTM) is based on the approach described in (30). This methodology was used to model relatively simple particle systems of practical interest, such as they occur in microfluidics, particle filtration, aerosols and blood flow problems in arteries (29; 30). Computation of the particle trajectory means the determination of particle position coordinates $x_p(t)$, $y_p(t)$ and $z_p(t)$ at each moment in time, t. This would require knowledge of particle velocity $\mathbf{v} = [v_x, v_x, v_x]$ so that the system of differential equations to be solved for each particle is:

$$\frac{dx_p}{dt} = v_x , \quad \frac{dy_p}{dt} = v_y , \quad \frac{dz_p}{dt} = v_z \qquad \text{[Eq. 5]}$$

In the Newtonian formalism, the velocity **v** depends on the acceleration created by the resultant of several forces (with components F_x, F_y, F_z) acting locally on the particle with mass m_p:

$$\frac{dv_x}{dt} = \frac{F_x}{m_p}, \quad \frac{dv_y}{dt} = \frac{F_y}{m_p}, \quad \frac{dv_z}{dt} = \frac{F_y}{m_p} \qquad \text{[Eq. 6]}$$

Most important is the drag force imposed by the fluid flow on the particle (30). The drag force is a function of particle size and density, liquid viscosity and relative velocity between particle and fluid. The relative importance of the forces involved in particulate deposition in a specific system is determined by the value of the Stokes number (30):

$$St = \frac{\rho_p d_p^2 u_{in}}{18\eta L} \qquad \text{[Eq. 7]}$$

where u_{in} and L are the imposed cross flow velocity and length of the channel, respectively, d_p and ρ_p are the particle diameter and density and η is the dynamic viscosity of the liquid. For very small values of the Stokes number ($St<10^{-5}$) and assuming no acceleration in the particle's motion during a given short time step, the particles follow the fluid streamlines (30; 32) and the error in tracking accuracy is below 1% (36). In the studied system, the calculated Stokes number was in the order of 10^{-6}, which means that the particle motion is convectively driven by the liquid flow. Other forces, such as inertia, gravity, lift, Brownian motion and electrostatic interactions are estimated to be negligible (30). The model considers particles substantially larger (2 μm) than the characteristic length scale of van der Waals forces (vdW), which can be estimated between ~10 nm and ~100 nm (29). Therefore, the inclusion of the DVLO potentials (the summation between vdW attraction forces and double layer repulsion forces) is not necessary for our model with particles in the micrometer size range. Particles are also sufficiently large (>500 nm) to neglect Brownian motion (30). Gravity is also not significant for the simulated time scale since the particles are almost neutrally buoyant.

Given the above assumptions, in this model we use a Lagrangian formalism, which does not include particle acceleration so that the position is only influenced by the local values of the fluid velocity. In this case, the position of a particle at time $t+\Delta t$ can be determined from its previous position at time t and the fluid velocity $\mathbf{u}$:

$$\frac{dx_p}{dt} = u_x, \quad \frac{dy_p}{dt} = u_y, \quad \frac{dz_p}{dt} = u_z \qquad [\text{Eq. 8}]$$

At each moment in time, the computational algorithm updates the particle position and checks for proximity to a wall. If particles are within 5 μm distance from a wall (spacer, membrane or glass) it was assumed that the particles can stick to that wall. All such determined collisions were counted as favorable deposition events. Because in reality not all collisions are leading to deposition, the time frame in the model is not comparable with the actual time.

Table 3: Model parameters

PARAMETER	DESCRIPTION	VALUE	UNITS	SOURCE
SYSTEM DIMENSIONS (for all dimensions see Figure 8)				
- Four-Tipped Star				
$L_x = L_y$	Length and width of the computational domain	1.17	mm	Measured See Figure 8
L_z	Height of the microstructure	0.208	mm	Measured See Figure 8
ε_L	Microstructured channel porosity	0.94	m³ liquid /m³ module	Calculated
- Teardrop				
L_x	Length along the main flow direction	1.409	mm	Measured See Figure 8
L_y	Width across the channel	1.384	mm	Measured See Figure 8
L_z	Height of the microstructure	0.208	mm	Measured See Figure 8
ε_L	Microstructured channel porosity	0.97	m³ liquid /m³ module	Calculated
- Commercial Spacer				
L_x	Length along the main flow direction	3.94	mm	Measured See Figure 8
L_y	Width across the channel	3.94	mm	Measured See Figure 8
L_z	Height between the membranes	0.711	mm	Measured See Figure 8
ε_L	Feed spacer channel porosity	0.89	m³ liquid /m³ module	Calculated
HYDRODYNAMICS				
ρ	Water density	1000	kg/m³	-
η	Water dynamic viscosity	0.001	Pa·s	-
u_{in}	Average water inlet velocity	0.14	m/s	Set value
PARTICLES				
d_p	Particle diameter	1.9	μm	From manufacturer
ρ_p	Particle density	1055	kg/m³	From manufacturer
NUMERICAL				
N_p	Number of particles	10^4	-	Chosen
Δt	Time step for particle trajectory calculation	10^{-4}	s	Chosen

Model Solution

The model was solved by combining MATLAB code (MATLAB 2010b, Mathworks, Natick, MA, *www.mathworks.com*) with the finite element method implemented in COMSOL Multiphysics (COMSOL 4.2a, Comsol Inc, Burlington, *www.comsol.com*) (17).

(1) A sequential approach has been used. Firstly a COMSOL model is used to calculate the liquid flow field, **u**, according to the following steps:

a) Define the model input parameters (see Table 3)
b) Construct the model geometry, define hydrodynamics and assign boundary conditions (Figure 9 - A).
c) Create a 3d mesh for finite element solution of the hydrodynamics. We used a mesh of ~3,000,000 tetrahedral elements with a maximum element size of 1.23×10^{-5} m in the volume close to the walls and a maximum size of 1.98×10^{-5} for the remaining volume. A detail of typical mesh used is shown in Figure 9-B. P1-P1 linear Lagrange mesh elements are used. This gives a number of degrees of freedom to be solved for of ~2,500,000 equations.
d) Solve the Navier-Stokes equations (Equations 3 and 4). The relatively fine discretization yields sufficiently accurate results with a reasonable computational effort (<6 hours per flow field).

(2) The solution provides the 3d distributions of velocity $\boldsymbol{u} = (u_x, u_y, u_z)$ and liquid pressure p. In addition, the viscous shear stress on the walls can be derived from the flow field. The flow velocity data was exported from finite element solution in COMSOL to a three-dimensional Cartesian grid in MATLAB.

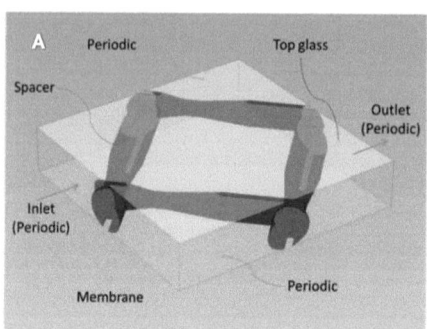
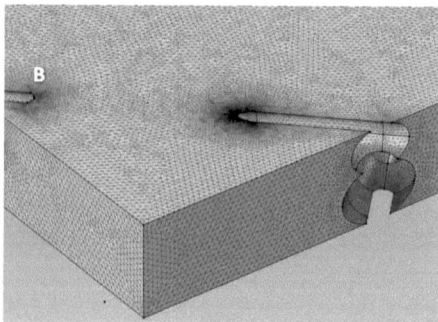

Figure 9: A - Computational domain of the realistic model spacer element. Membrane on the bottom side, glass on the top. Liquid inlet, outlet and lateral are periodic boundaries. B - Detail of spacer element with the finite element mesh used.

A matrix of predetermined resolution $\Delta x = \Delta y = 9.7$ μm and $\Delta z = 7.7$ μm. This resolution implied transferring matrices of maximum size of 400×400×100 elements.

(3) A MATLAB script solves the ordinary differential equations of particles trajectories for a set number of particles within a time period. The liquid velocity **u** is used to support the convective transport of the particles. The particle position is computed from a forward Euler discretization of the following equation:

$$x_p(t + \Delta t) = x_p(t) + u_x(\Delta t), \; x_p(t + \Delta t) = x_p(t) + u_x(\Delta t), \; x_p(t + \Delta t) = x_p(t) + u_x(\Delta t) \; \text{[Eq. 9]}$$

where the particle velocity $v_{x,y,z}$ equals the liquid velocity $u_{x,y,z}$ corresponding to the current position $x_p(t), y_p(t), z_p(t)$. The local velocities were calculated by tri-linear interpolation from the 3d grid solution imported from COMSOL. Particle trajectory calculation is performed with a time step $\Delta t = 10^{-4}$ s. Smaller time steps did not seem to affect the results anymore.

Particles are always sent into the fluid from a point with coordinates (x_0, y_0, z_0) situated on the inlet boundary (at $x=0$). The initial positions are chosen randomly, but proportionally to the liquid velocity (i.e., the higher the liquid velocity, the higher chance to send a particle from that point).

For each particle, the algorithm checks for proximity to a wall and particles within 5 μm distance from a wall (spacer, membrane or glass) are assumed to be stuck. The stuck particles are immediately replaced with new particles injected in the inlet. Particles exiting the domain through the outlet boundary ($x=L_x$) are replaced as well. Lateral boundaries ($y=0$ and $y=L_y$) are periodically connected so that a particle exiting through one boundary will re-enter through the opposite surface. In this way, a number of $N_p=10000$ particles are always present within the system.

For visualization of attached particles, different colors are used function of their attachment place: yellow on membrane, red on feed spacer and green on glass.

4. RESULTS AND DISCUSSION

3.1 MEMBRANE CHARACTERIZATION

The effect of non-Solvent

During the phase separation process, exchange of solvent and non-solvent takes place between the polymer solution film and the coagulation bath (10; 13; 37). Both horizontal and vertical shrinkage occur, strongly influenced by the ratio between non-solvent in-flux and solvent out-flux from the polymer solution (10). Replication of several polymer solutions onto one silica wafer with two shapes built-in was investigated, to evaluate the influence of polymer composition and coagulation bath on the final structure and surface properties.

In this section, differences in the cross section morphology for a set of two microstructured supports have been investigated, when diverse coagulation baths are employed. For instance Figure 10 shows that when water is used as a coagulation bath, macrovoids are present inside the features and they are bigger than the macrovoids present in layer underneath. In this regard Bikel et al., have showed that microstructured supports are usually a location populated by macrovoids (10). The formation of macrovoids is principally due to no constant composition at the interface between polymer solution and coagulation bath (38). Experimentally it has been shown that the formation of macrovoids in membranes prepared from a polymer/solvent/non-solvent system can be suppressed in many ways. The simplest way is the addition of more solvent into the coagulation bath (10; 13; 39). No macrovoids formation it has been observed when NMP concentration in the coagulation bath exceeded 75% (10). However, these membranes were not very well solidified, therefore a second coagulation bath of only tap water was necessary. To avoid this further passage in this work a coagulation bath composed by a mixture of water/NMP (50% v/v) was used.

Figure 10: *Cross section of a membrane obtained from a 15% PES solution in NMP, coagulated in water. Larger macrovoids are inside the features than in the layer in membrane.*

SEM pictures in Figure 11 show that the membranes prepared with PES (15% wt.) coagulated in tap water presents a morphology influenced from the thickness. Macrovoids become bigger in the meanwhile the thickness increases. On the contrary, membranes coagulated in the mixture composed by 50% water/NMP, present a morphology independent from the thickness. Furthermore, as expected, the morphology of the layer was influenced by the coagulation bath employed. When tap water is employed as non-solvent, the overall porosity is higher. This is explicable since the ratio between non-solvent in-flux and solvent out-flux from the polymer solutions during coagulation is higher when water is used instead of the mixture water/NMP. The micrographs clearly show for the PES membranes coagulated in tap water a regular and "lighting-like" larger macrovoids, departing immediately after a very thin layer as a surface. On the other hand for the membranes coagulated with the mixture water/NMP show a thick, dense "sponge-like" top layer, followed by an another layer of less tight, "finger-like" porous as the direct consequence of the presence of NMP in the coagulation bath (40). Moreover, emphasizing on the pores morphology it can be seen that using tap water, leads into a lower pores size than the mixture of water plus solvent.

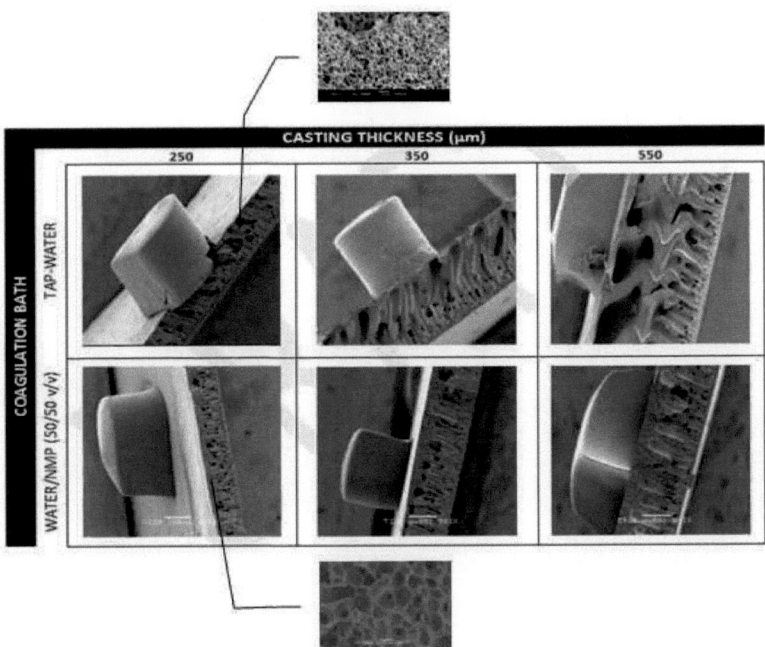

Figure 11: The effect of the casting thickness and coagulation bath on the final structure of the film. Macrovoids are less formed when solvent is used in the coagulation bath (bottom line). The pictures in the box were taken with ×180 magnification, the two pictures outside the box at ×6000.

A not negligible side-effect when casting at different thicknesses was the observed non-homogeneous thickness of the final layer in fabricated membranes. As a consequence of the manipulations that the silica mold with the polymer solution already cast undergoes, we noticed that the higher the casting knife was set, the higher was the inconsistency of the final thickness of the film. To prove this effect, solutions were prepared using PES dissolved in NMP. For each concentration, we have cast the polymer solution at different heights and for each concentration two coagulation baths have been used (see Table 1). Via linear regression fitting (R^2 was higher of 0.97 in all cases), the mean final thickness of the membranes (T_{cast}) versus casting thickness was obtained. The results are plotted in Figure 12 and resumed in Table 4.

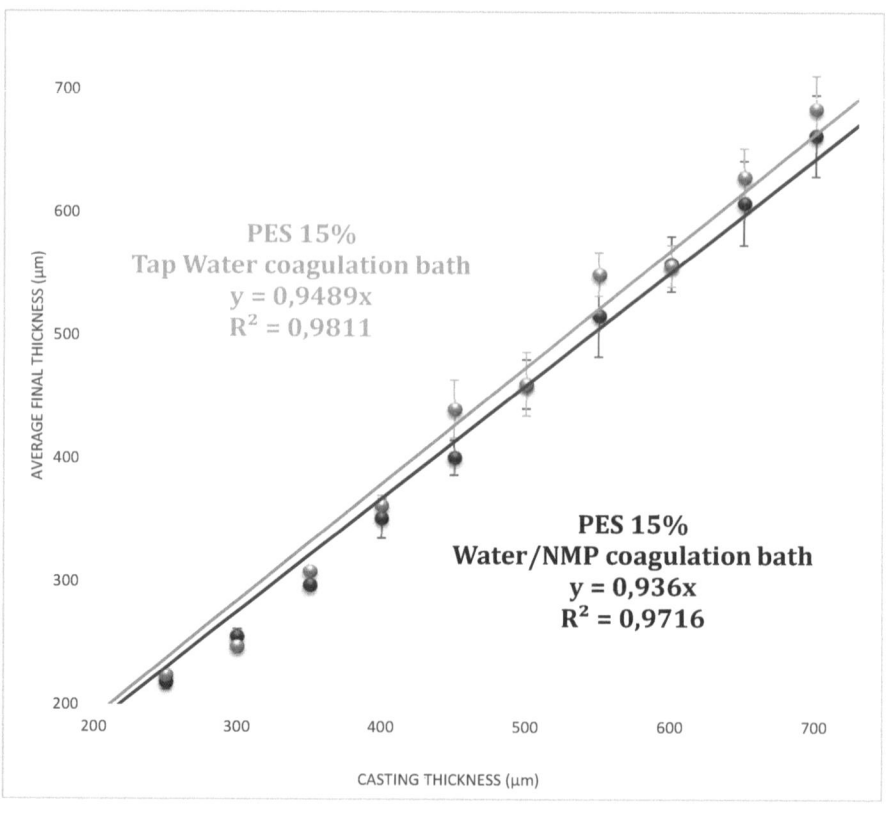

Figure 12: Plot of the final thickness (measured including the microstructures) versus the casting thickness for a solution containing 15% PES in NMP coagulated in water/NMP (50/50 v/v) bath and tap water bath.

Table 4: Comparison between casting thickness, the average final thickness and the average final height of the microstructures of the PES membranes obtained.

PES 15%(w/w)		
Casting thickness (μm)	Averaged final total thickness *	Averaged Height ± sd ** of the Microstructures ***
250	218.5	206.7 ± 8.5
300	255.6	207.1 ± 7.9
350	297.3	207.4 ± 7.5
400	351.6	207.5 ± 7.5
450	401.0	207.2 ± 8.0
500	460.7	206.5 ± 7.3
550	516.5	208.4 ± 7.2
600	558.5	207.0 ± 7.3
650	608.0	206.8 ± 7.4
700	662.8	208.5 ± 6.9

* Each membrane was measured 10 times in different spots and the average value was taken.
All the values are in micrometers
** Standard deviation.
*** Measurements have been taken from SEM images with the software SEMafore.

The entity of the vertical shrinkage has been calculated using the following equation (10):

$$T_{cast} = \frac{d_{end}}{d_{cast}} = \frac{d_{cast} - d_{shrinkage}}{d_{cast}} = 1 - S_k \qquad \text{[Eq. 10]}$$

where d_{end} is the final thickness of the overlying film, d_{cast} is the casting thickness and S_k is the shrinkage. The thickness measurements were taken using a digital micrometer - Mitutoyo (0 – 25 mm) and the microstructures heights measured with SEMafore are grouped in Table 4. The data showed in Figure 12 indicates a decreasing accuracy of the final film when the casting thickness increases. A cross section micrograph of a 700 μm cast of a PES 15% membrane coagulated in water/NMP (Figure 13) clearly confirms the discrepancy in height. On the contrary, microstructures are very well replicated following the same shrinkage trend, independent from the thickness cast. The overall shrinkage is quantified to be between 13 and 20%. For the range of casting thickness studied, using different coagulation baths therefore does not reduce the problems related to the fabrication method described in this thesis. Over 400 microns, the error bars, calculated as the relative standard deviation, tend to increase (for membrane cast at 700 μm reached 34%), which can be basically translated in non-regular film layer. Pore size and morphology were uniform all over the cross-section, independent of the thickness of the membranes, but strongly influenced on the coagulation bath employed.

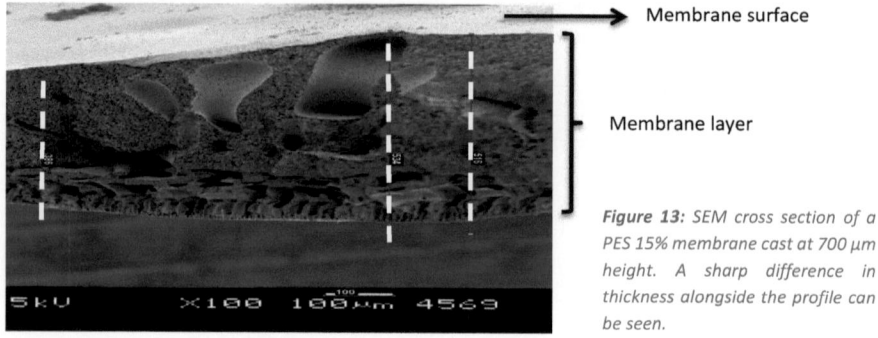

Membrane surface

Membrane layer

Figure 13: SEM cross section of a PES 15% membrane cast at 700 μm height. A sharp difference in thickness alongside the profile can be seen.

The Effect of PVP

Polyvinylpyrrolidone (PVP) is generally added in order to increase permeability, giving more hydrophilicity to the membrane, with no effects on selectivity (11; 37). More hydrophilicity is given because of the interaction between N–C=O group in PVP and the O=S=O from PES and could cause an increase in the big pores fraction leading as said to higher permeability (40). In Figure 14 and Figure 15 SEM pictures (left-hand side) together with the AFM scans (right-hand side) of the six membrane types studied are presented.

A small round piece of each membrane was cut and in the first place analyzed via AFM. Subsequently the same sample was placed in the electronic microscope.

For AFM micrographs the area scan size is 625 μm^2 (25×25) in all cases. The values of R_a and R_q for the six types of membranes studied are summarized in Table 5.

Table 5: The average roughness R_a and the average of the height deviation R_q

Membrane Composition	Coagulation Bath	R_a *	R_q *
PES	Tap Water	26.0	36.6
PES/PVP(K12)	Tap Water	28.1	37.4
PES/PVP(K30)	Tap Water	47.6	65.1
PES	Tap Water/NMP (50% v/v)	50.7	65.1
PES/PVP(K12)	Tap Water/NMP (50% v/v)	58.7	74.1
PES/PVP(K30)	Tap Water/NMP (50% v/v)	77.4	101.0

It is clear that according to Figure 14, the membranes prepared with PES and PES+PVP have different topography with a substantial difference in surface porosity in spite of the concentration of the additive is low. Particular relevance has the PVP(K30) picture (Figure 14-C); it shows a sharp increasing number of pores. It should be noted that roughness increases when PVP is added and more for longer PVP chains (showed in Table 5).

Figure 15 shows the other three types of membranes, this time coagulated in water/NMP mixture. The three pictures confirmed what we saw before. PVP increases the overall porosity of the surface and the roughness values as well (see Table 5) and there are no sharp differences caused by the utilization of different coagulation baths.

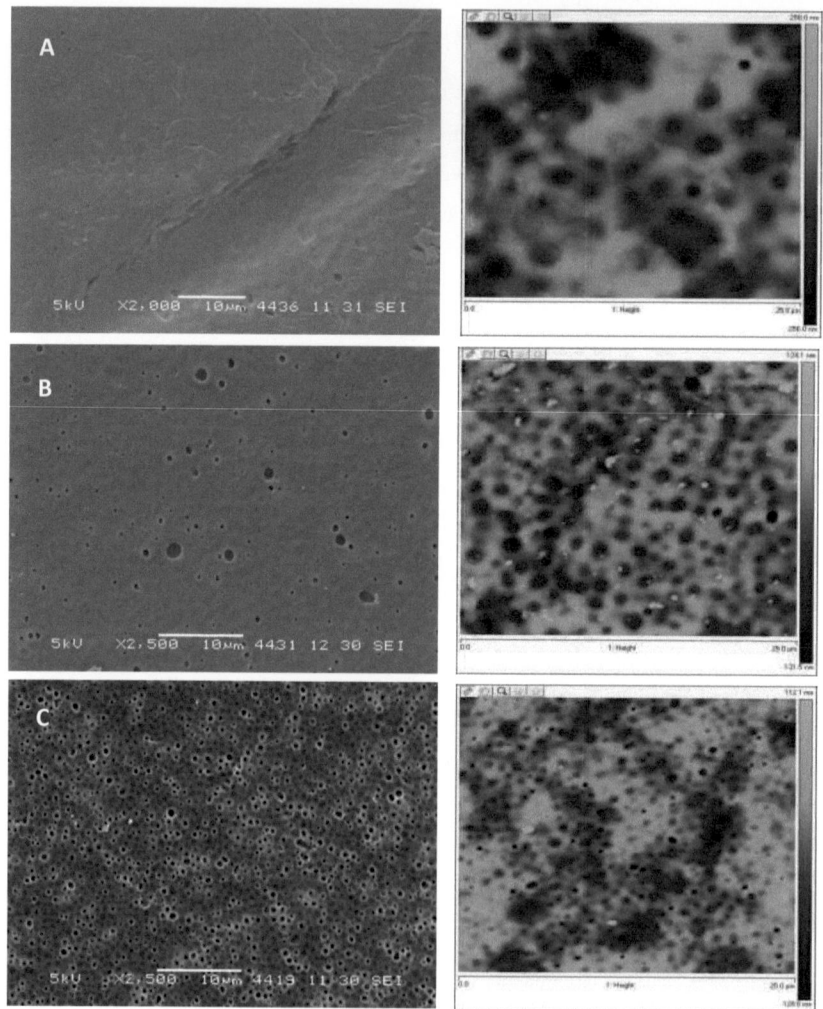

Figure 14: SEM and two-dimensional AFM pictures of three membrane types studied. A)PES 15%; B)PES+PVPK12; C)PES+PVPK30, all coagulated in water bath.

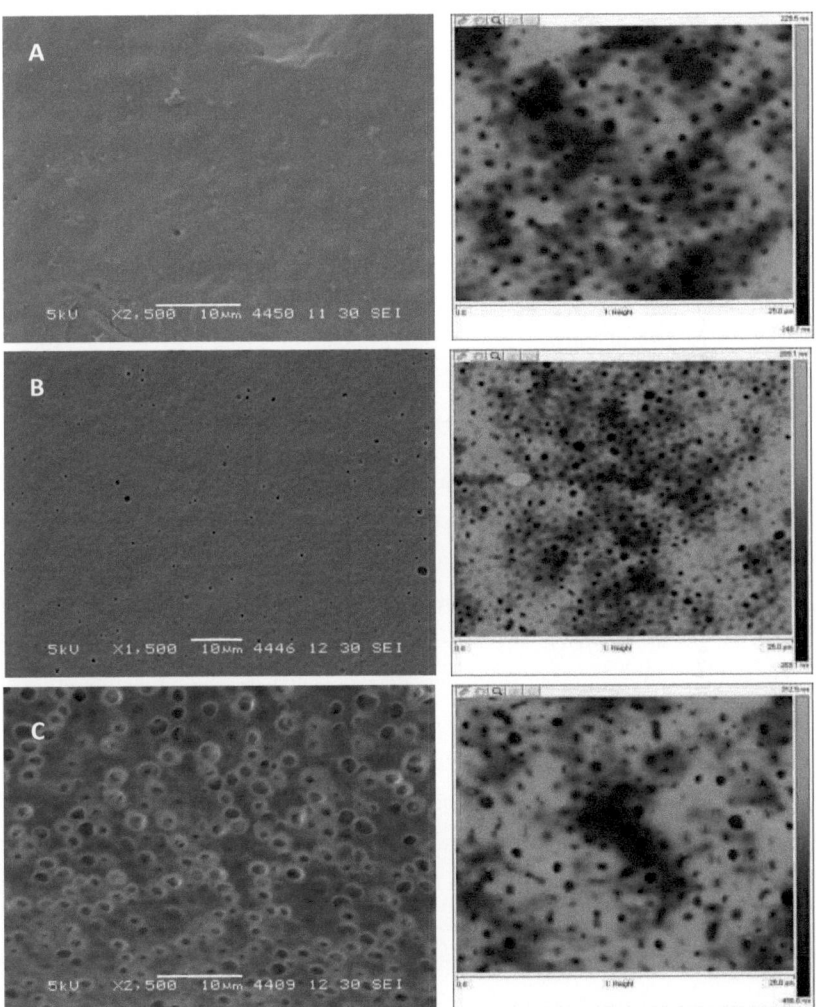

Figure 15: SEM and two-dimensional AFM pictures of three membrane types studied. A)PES 15%; B)PES+PVPK12; C)PES+PVPK30, all coagulated in water/NMP bath.

3.2 PARTICLE DEPOSITION

The commercial spacer and two different microstructured membranes were placed in the Membrane Fouling Simulator operated under cross-flow conditions without permeation. The flow and particle deposition were then studied numerically for similar spacer geometry. Based on the experimental and simulation results obtained for all the membrane modules, the effect of spacer and microstructures orientation on the flow pattern and particle deposition is discussed in the following sub-chapters.

Particle attachment in time

The development of particle deposition patterns in a membrane channel with commercial feed spacer (DOW® chemicals) as a function of time has been investigated. Figure 16 gathers pictures of the same representative feed spacer element at different times, from the start with clean spacer up to 24h of particle suspension circulation. From the several experiments performed during the thesis work, only one case will be presented here. The mean flow was sent at 45° angle on the filaments so that the average cross-flow velocity was 0.14 m/s.

At the start of the experiment (t=0), the spacer and membrane were clean (particle free), as shown in Figure 15. After only 2 hours, traces of particles deposited are already visible. The feed spacer seems to be the first to attract the particles traveling across the channel. When micron-range particles are suspended in a liquid stream, they are subjected to several forces: drag, lift, inertia, gravity, Brownian. Additionally, interactions with the walls may take place, such as attraction and repulsion (Van der Waals, Lennard-Jones) function of material properties of particles and walls. When the liquid velocity is very slow (i.e., near the walls) the contribution of particle-wall forces and of gravity and Brownian motion may become relevant and justify the particle attachment. We believe that the particles' preference for the spacer instead of membrane/glass is related to material properties. After five hours two other deposition areas can be distinguished. The first one, circled in the pictures with the yellow color, is on the membrane just in front of the entrance through the "tunnel" below the spacer fibers. The second area (marked with a green circle) is appearing on the top glass plate, symmetrically with the first area in respect to flow axis, also in front of the entrance to the bottleneck created by fibers. This deposition pattern indicates that the deposition on the membrane (bottom surface) is not the effect of gravity, because a similar pattern appears on the top glass overcoming the gravity force.

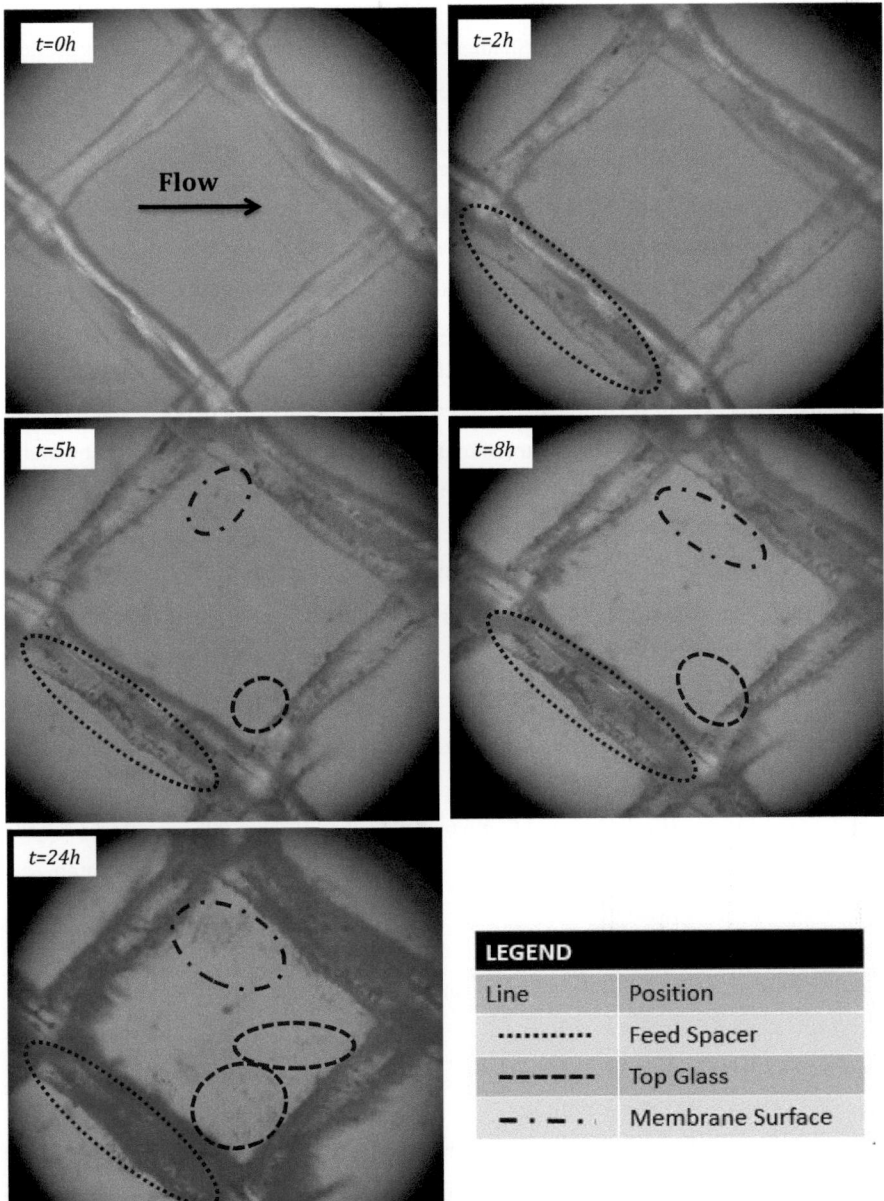

Figure 16: *Evolution of particles deposition patterns as a function of time in the Membrane Fouling Simulator. All the pictures are for the same spacer element. Flow is from left to right in all cases. Circled areas with different line styles indicate distinct regions of attachment: spacer fibers; glass (top surface); membrane (bottom surface).*

Both the areas on the membrane and the one on the glass are situated in the vicinity of the bottleneck between the membrane/glass and the feed spacer. The presence of such flow bottlenecks is due to variations of spacer filament thickness. As shown in the following section, this behavior can be theoretically explained by the numerical model. The deposition on membrane and glass is not identical though. This can be caused by: (1) different material properties (so that particles are attached more to one material or to the other) and (2) differences in filament thickness (so that the bottlenecks are not exactly the same width). In this experiment, the thinner spacer filament was in contact with the membrane surface so the fluid was preferentially lifted up in order to overcome the obstacle - which led to deposition the glass. *Vice versa* is the situation for the yellow marked spot on the membrane. The picture of the fouling after 8 hours clearly shows more particles attached in all described areas. After 24 hours the spacer fibers appear completely covered with red particles, with the development of several filamentous formations. The nature of these filaments is still not understood and under investigation. A potential cause for the formation of such filaments is particle agglomeration. The only zones not covered with particles are the junctions (crossings) of the spacer filaments, indicating that in that region there is no flow. This is a confirmation that the feed spacer was perfectly pushing against the membrane and the glass and that the main supporting areas were these junctions.

The effect of feed spacer geometry

The effect of shape and orientation of feed spacer on particle deposition was evaluated by performing fouling experiments and comparing the results with 3d numerical simulations. Commercial feed spacers (Dow Chemical Company) placed at 45° and 90° in respect with the main flow direction and their upside-down corresponding configurations were evaluated:
- Configuration 1 has the spacer at 45° ("diamond"), with the thicker spacer filaments in contact with the membrane.
- In a similar Configuration 2 ("diamond") the spacer is placed upside-down, with the thinner filaments touching the membrane.
- Configuration 3 has the spacer at 90° ("ladder") with filaments parallel with the main flow direction (x) in contact with the membrane.
- Configuration 4 is also "ladder" oriented, but rotated such that filaments perpendicular to the main flow direction (x) are touching the membrane.

Additionally, microstructures, with shapes of four-tipped stars (Configuration 5) and teardrops in two different orientations (Configuration 6 and 7) were tested. Experimentally, for all configurations tested, specific and reproducible patterns in all spacer array elements could be observed, as shown in Figure 17(A, B, C) Therefore, for results analysis and for further model simulations we focused on a single representative spacer element from the whole array. Detailed simulation results and experimentally observed patterns in a single spacer element are shown in Figure 18.

Commercial net-shaped spacers

The particle deposition patterns obtained experimentally for two net-shaped spacer configurations (Configuration 1 and 2) compared well with the model results (Figure 18). Deposition on the spacer can be observed along the whole length of the filament (areas marked by red contours). The deposition on membrane and glass (yellow and green contours, respectively) is mainly concentrated upstream the bottlenecks created by spacer fibers. The specific deposition zones on the top glass for both configurations indicate that the gravity does not have a significant influence on the particle deposition patterns. In configuration 1, at u_{in}=0.14 m/s, there is also a middle stripe deposit on the membrane. This area seems to correspond to a zone of high shear stress (shown as light gray shades in Figure 18). However, because particles also attached in other areas (e.g., at lower shear), we cannot certainly correlate the shear value with the amount of particle deposited. The position of preferential regions for particle deposition in early stages (<24 hours) observed experimentally is qualitatively similar to the patterns resulting from the 3d model.

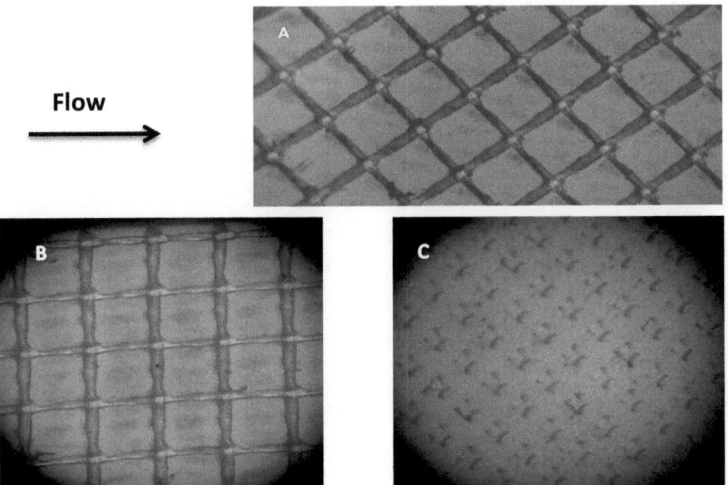

Flow

Figure 17 *Particles fouled membrane coupons with different spacer orientation (A and B - commercial DOW spacer. C - four tipped star)- A specific and reproducible pattern can be seen across the length of the channel. Pictures have been taken after 18 hours of experiment under cross flow. Flow direction is from left to right.*

CONFIGURATION 1

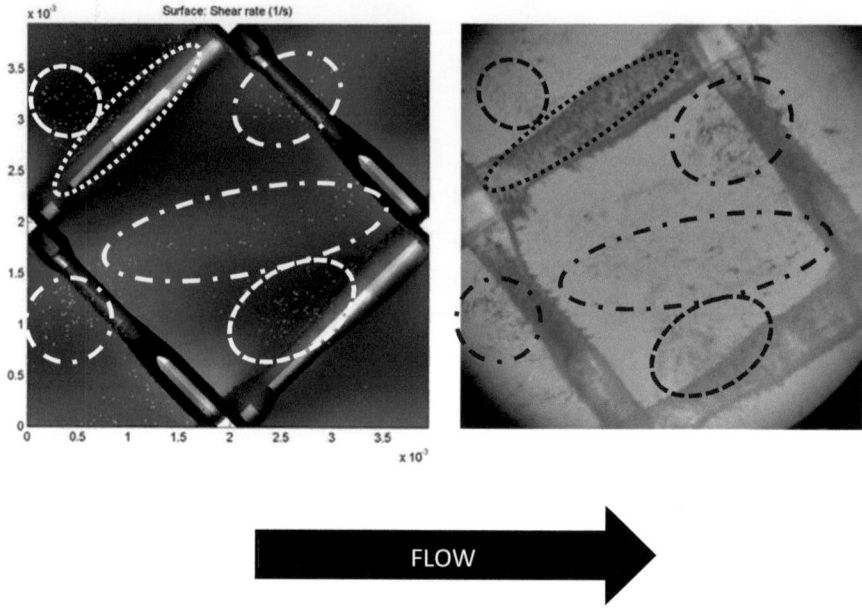

FLOW

CONFIGURATION 2

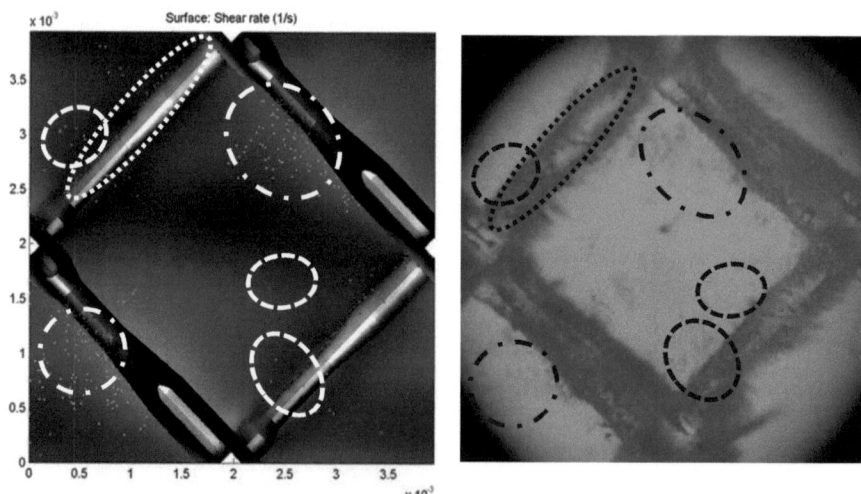

Figure 18: Compared particle deposition patterns in different spacer configurations. Numerical model results (left) and in-situ visual observation of the feed spacer channels (right). For the model results, the shear stress on the membrane and spacer is shown with a gradient of gray (light gray = high shear). The deposition regions are highlighted with contours of different colors: green=top glass; red=spacer; yellow=membrane.

CONFIGURATION 3 AND 4

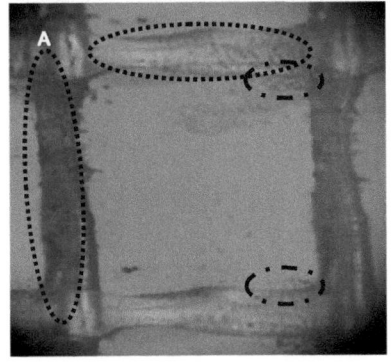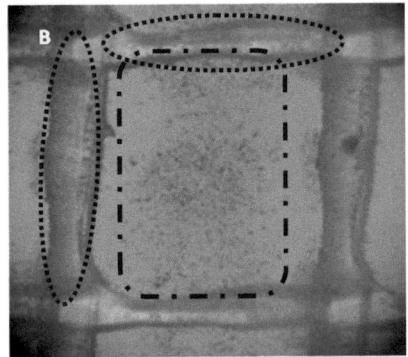

Figure 19: *Particle deposition with spacer orientation perpendicular respect to the flow. Flow is from left to right. The deposition regions are highlighted with contours of different colors: green=top glass; red=spacer; yellow=membrane.*

The two other two possible orientations of the commercial feed spacer have been investigated (Configurations 3 and 4 with "ladder" oriented spacer) and shown in Figure 19 after 18 hours of continuous flow. The difference in the position of the two plastic filaments influences much more the membrane fouling pattern. When the filament in contact with the membrane is perpendicular to the flow, the membrane surface becomes extremely fouled. Most of the particles deposit uniformly in the middle area between two filaments (Figure 19B). On the contrary, when fibers touching the membrane are parallel with the main flow the membrane surface appears rather clean, with particles accumulation only in the two corners upstream the filament nodes. Remarkably, not much deposition occurred on the spacer fibers parallel with the flow, especially in Configuration 3. Further studies, coupling experimental results with numerical model predictions will be implemented to better understand this situation.

Microstructured spacers

In microstructured Configurations 5, 6 and 7, the deposition of particles upstream the structures regardless of the orientation is presented in Figure 20. The observation of particulate fouling on the structures, confirms that the particles are convectively forced onto the front surface leading to the formation of deposits in these areas. Preferential deposition in front of the obstacles was also reported experimentally by Ngene (9). For microstructures, a correlation between shear stress and particle attachment can be observed, as suggested also by Ngene (9).

Flow ⟶

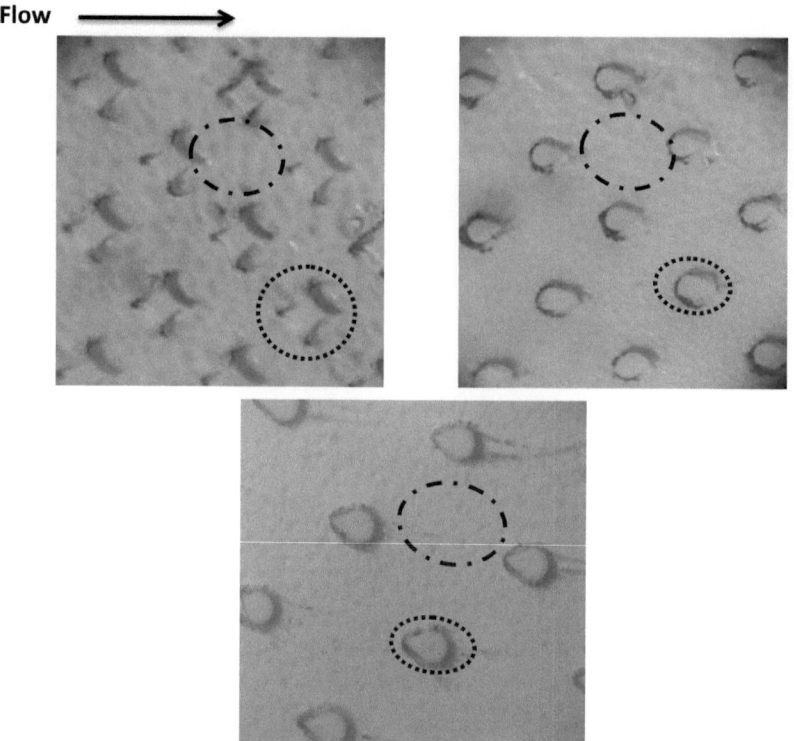

Figure 20: *Compared particle deposition patterns in different microstructures shapes configurations. The deposition regions are highlighted with contours of different colors: green=top glass; red=spacer; yellow=membrane.*

CONFIGURATION 5, 6 AND 7

The lowest shear rate areas are situated behind the structure (downstream of the obstacle) and the highest shear rate areas are on the sides or in front of the obstacles. On the other hand, because of high shear there is a limited mass of particles that can be deposited on the surface of the obstacles. High shear is correlated with high particle transport to the surface (thus more particles can attach), but at the same time, high shear may also lead to more detachment of particles from the structure surface.

Interestingly, for all microstructures, there is no observable deposition on the membrane or glass surfaces. This can be due to the quasi-two-dimensional (i.e., with mainly x and y velocity components) flow pattern in microstructured channels, as explained in the further Section "Correlation between flow field and particle deposition".

The effect of the liquid velocity

The influence of the fluid velocity on particle deposition has also been investigated, only for the Configuration 1. With the flow rate reduced to approximately one third (u_{in}=0.05 m/s) the pattern obtained is quite different form the one resulted when the cross flow velocity was higher (u_{in}=0.14 m/s). Again, the feed spacer collected most of the traveling particles. At low velocity, there is a more uniform particle distribution along the whole spacer length between crossings, whereas at high velocity the particles are concentrated in the bottleneck area. This could be correlated with a more uniform flow distribution at low velocities, while at high velocities more jet-like streams develop in the bottlenecks (Figure 21). The difference however is larger in deposition patterns on the membrane. In Figure 21 B, as discussed before, two regions of attachment on membrane can be distinguished at high velocity, while at reduced flow rate one larger deposition area is covering half of the surface of the square element. Apparently, at low velocity the deposition area is more widespread. The pattern on the glass seems to be so much influenced by the fluid velocity.

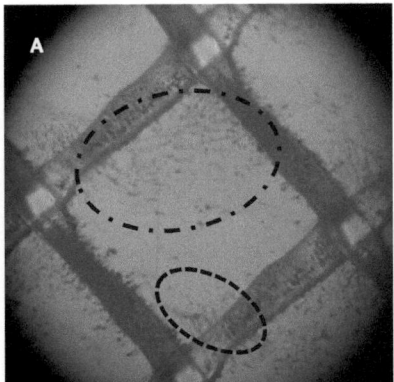 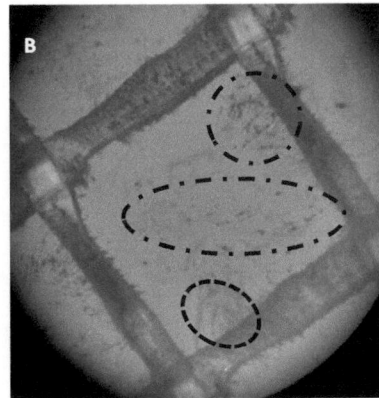

Figure 21: Particles deposit pattern under two different cross flow velocity. Flow rate is from left to right. A) Cross flow velocity is 0.05m/s. B) Cross flow velocity is 0.14 m/s.

Correlation between flow field and particle deposition

Even if in most of the cases the simulated particle deposition patterns are at least qualitatively similar to the practically observed ones, the question still remains what could be the main hydrodynamic parameter or effect that drives the preferential particle deposition in certain regions. Based on the computed three-dimensional flow fields, we attempt here an interpretation of their influence on particle attachment. The local values of two flow variables were identified as having a major influence: the vertical (z-component) of the flow velocity (u_z) and the shear rate τ.

The complex geometry of the net-shaped feed spacer creates an intricate flow pattern, influenced also by the orientation and the position of the filament in respect to the flow. The flow field is fully three-dimensional and follows a tortuous path up and down past the spacer filaments (direction z). On the other hand, a quasi-two-dimensional flow pattern has been observed for microstructured membranes (not shown), with (at least theoretically) no flow in the z direction. It appears that the spatial distribution of this z velocity component (u_z) is important for determining the deposition pattern. For microstructures, no deposition on the membrane surface was observed, possibly correlated with the absence of vertical velocities (Figure 22). For net-shaped spacers, the high attachment areas correspond to regions of high u_z as seen in Figure 22 (A, B, C) in a section through the xy plane in the middle of the feed channel (at height $z = 356\ \mu m$). In Configuration 1 (Figure 16A) the highest u_z in absolute values are

Configuration 1   Configuration 2

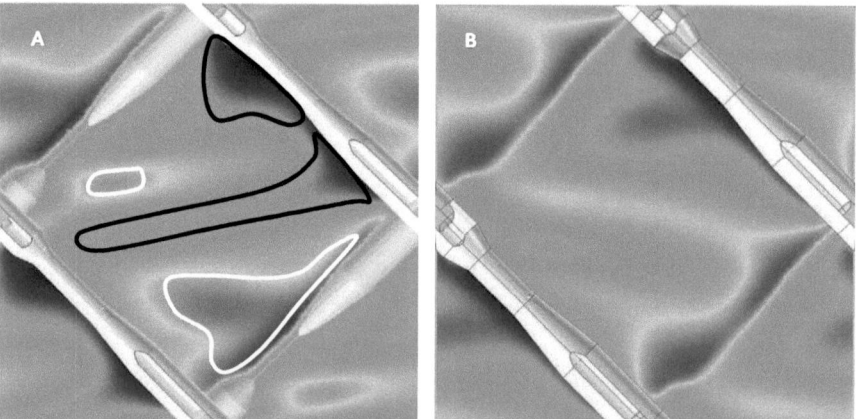

Figure 22: z component liquid velocity distribution in the feed channel of one element. Two dimensional distributions of liquid velocity are presented in the middle plane in the z direction for normal spacer in two different configurations (A and B). Black contour is z velocity pushing against the membrane; white contour is z velocity pushing against the glass. Flow is from left to right in all cases.

just upstream the filaments. Two more areas of high absolute u_z can be distinguished in the middle of the spacer element, along the x direction. Consequently, for high negative u_z (flow pushed "down" – black contour in Figure 22 A, B) the particles attach to the membrane, whereas for high positive u_z (flow pushed "up" - white contour in Figure 22 A, B) we found particles attached to the glass plate. For Configuration 2 (Figure 22 B) the spacer position is flipped and the preferential attachment zones on glass and membrane are accordingly reversed. In other words, these two configurations are the reproduction of an actual situation of a spiral wound element, in which the two membranes are separated by the feed spacer.

For the microstructures, because u_z is theoretically absent, we did not find relevant deposition on the glass and on the membrane. The main deposition is actually at areas of high shear rate situated upstream the obstacles, regardless their orientation and shape. The star shaped structure presents a recirculation zone behind the structures under the conditions tested. This is not the case for the tear drop, which have no recirculation zones downstream of the structures under the flow conditions regardless to the orientation (9).

5. CONCLUSIONS AND RECOMMENDATIONS

(1) Microstructured membranes with two different shapes of obstacles were prepared. The new method proposed here is much faster than the one described by Ngene et al. (9) and a perfect replication of the features has been achieved.

- The prepared membranes have to be thinner than the MFS height (780 μm) because the optimal height for casting these membranes is 350 μm. Therefore, a support layer has to be put in the MFS beneath the membrane in order fill this gap.

- From SEM and AFM images it appears that surface topography is qualitatively similar for the two pair of membranes studied (PES with and without PVP) using two different coagulation baths. Nevertheless, big differences can be seen when two different molecular weight PVPs are added. The AFM gives better and more detailed surface profiles than the SEM. The addition of PVP changes mainly the surface pore number, reflected in increasing membrane roughness, in proportion of the PVP molecular weight.

- Difference in composition of coagulation baths does not have visible effect on the surface properties but rather on the final structure of the membranes. More macrovoids are present when only water is used.

(2) A series of experiments were carried out to study the deposition patterns of polystyrene microspheres (mimicking microbial cells) in membrane channels with spacers. A three-dimensional numerical model for particle deposition driven by liquid flow was also developed.

- The flow pattern proved to be the main factor responsible for the characteristic particle deposition areas observed experimentally. In general, particles strongly attach to the spacer and microstructures regardless to their orientation. However, the geometry and the orientation of the spacer do influence the membrane fouling behavior. Several typical zones prone to membrane fouling (through particle deposition) were identified.

- The fluid velocity component towards the membrane (the z component) seems to have particular importance for the initiation of particle deposition. Due to the truly three-dimensional flow pattern, the commercial net-shaped spacers promote more particle deposition to the membranes than the microstructured ones, which allow a quasi-two-dimensional flow. Because the net-shaped spacer, glass and membrane have different surface properties, differences in particle deposition on these surfaces have also been revealed.

The implementation of free standing microstructures instead of normal net-shaped spacer can be a good solution for the future in membrane operations because it seems to be less prone to fouling. However, the difficulties in reproducibility during fabrication and characterization (thickness is the principal issue to overcome) represent the biggest challenge to face in order to find a future and realistic application of these new membranes concept. Moreover, it is not clear whether the mass transfer needed to overcome concentration polarization is sufficiently enhanced by these microstructures.

On the computational aspects, it is desirable to extend the current model with other representative forces (mainly particle-particle interactions), to other spacer configurations and different operational conditions. Especially, the model should be improved to better describe the particle deposition in the channels with net-shaped spacers in ladder configuration. To extend the method to such problems, further research is necessary for development of a model that can account for the effects of fluid forces on the all possible interactions between particles, (i.e. during particle collisions), while still retaining a high efficiency of the computational method.

BIBLIOGRAPHY

1. http://science.howstuffworks.com/environmental/earth/geophysics/question157.htm.

2. J.E. Drewes, P. Xu, D. Heil, G. Wang. *Multibeneficial Use of Produced Water Through High-Pressure Membrane Treatment and Capacitive Deionization Technology*. *Colorado*, U.S. Department of the Interior Bureau of Reclamation, 2005.

3. C.P. Koutsou, S.G. Yiantsios, A.J. Karabelas. *Direct numerical simulation of flow in spacer-filled channels: Effect of spacer geometrical characteristics*. Journal of Membrane Science, 2007, Vol. 291.

4. J.S Vrouwenvelder. *Biofouling of spiral wound membrane systems*. Delft, Delft University of Technology, 2009.

5. D. Dendukuri, S. K. Karode, A. Kumar. *Flow visualization through spacer filled channels by computational fluid dynamics-II: improved feed spacer designs*. Journal of Membrane Science, 2004, Vol. 249.

6. A.I. Radu, J.S. Vrouwenvelder, M.C.M. van Loosdrecht, C. Picioreanu.*Modeling the effect of biofilm formation on reverse osmosis performance: Flux, feed channel pressure drop and solute passage*. Journal of Membrane Science, 2010, Vol. 365.

7. C. Picioreanu, J.S. Vrouwenvelder, M.C.M. van Loosdrecht. *Three-dimensional modeling of biofouling and fluid dynamics in feed spacer channels of membrane devices*. Journal of Membrane Science, 2009, Vol. 345.

8. I. Ngene *Real time visual characterization of membrane fouling and cleaning*. Enschede, Twente University, 2010.

9. M. Bikel. *Phase separation microfabrication*. Enschede, Twente University, 2009.

10. J. de Jong, B. Ankone, R. G. H. Lammertink, M. Wessling. *New replication technique for the fabrication of thin polymeric microfluidic devices with tunable porosity*. Lab on a Chip, 2006.

11. M. Girones, I.J. Akbarsyah, W. Nijdam, C.J.M. van Rijn, H.V. Jansen, R.G.H. Lammertink, M. Wessling. *Polymeric microsieves produced by phase separation micromolding*. Journal of Membrane Science, 2006, Vol. 283.

12. P.Z. Culfaz. *Microstructured hollow fibers and microsieves: fabrication, characterization and filtration applications*. Enschede, Twente University, 2010.

13. A. Subramania, E.M.V. Hoek. *Direct observation of initial microbial deposition onto reverse osmosis and nanofiltration membranes*. Journal of Membrane Science, 2008, Vol. 319.

14. C.Y. Tang, Y.N. Kwon, J.O. Leckie. *Fouling of RO and NF Membranes by Humic Acid-Effect of Solution Composition and Hydrodynamic Conditions.* Journal of Membrane Science, 2007, Vol. 290.

15. Y.L. Li, K. L. Tung, Y. S. Chen, K. J. Hwang, K. L. Tung, Y. S. Chen, K. J. Hwang *CFD analysis of the initial stages of particle deposition in spiral-wound membrane modules.* Desalination, 2011, Vol. 287.

16. A.I. Radu, J.S. Vrouwenvelder, M.C.M. van Loosdrecht, C. Picioreanu. *Effect of Flow Velocity, Substrate Concentration and Hydraulic Cleaning on Biofouling of Reverse Osmosis Feed Channels.* Chemical Engineering Journal, 2012.

17. J.S. Vrouwenvelder, *The membrane fouling simulator as a new tool for biofouling control of spiral-wound membranes .* Desalination, 2007, Vol. 204.

18. C.Y. Tang, T.H. Chong, A.G. Fane. *Colloidal interactions and fouling of NF and RO membranes: a review.* Advaces in colloid and interface Science, 2011, Vol. 164.

19. R. Rusconi, S. Lecuyer, N. Autrusson, L. Guglielmini, H. Stone. *Secondary flow as a mechanism for the formation of biofilm streamers.* Biophysical Journal, 2011, Vol. 100.

20. C. Kleinstreuer, T.P. Chin. *Analysis of Multiple Particle Trajectories and Deposition Layer Growth in Porous Conduits.* Chem. Eng. Commun., 1984, Vol. 28.

21. V. Rajasekar. *Double Layer calculations for the attachment of a colloidal particle with a charged substrate.* Walpole (MA - USA) : Bird Machine Company, 1991.

22. C.Y. Tang, T.H. Chong, A.G. Fane. *Colloidal interactions and fouling of NF and RO membranes: A review.* Advances in Colloid and Interface Science, 2010, Vol. 164.

23. H.X. Zhu, M. Elimelech. *Colloidal Fouling of Reverse Osmosis Membranes: Measurements and Fouling mechanisms.* Environmental Science and Technology, 1997, Vol. 31.

24. W.S. Ang, M. Elimelech. *Protein (BSA) Fouling of Reverse Osmosis Membranes: Implications for Waste Water Reclamation.* Journal of Membrane Science, 2007, Vol. 296.

25. Y. Wang, F. Wicaksana, C.Y. Tang, A.G. Fane. *Direct Microscopic Observation of Forward Osmosis Membrane Fouling.* Science Technology, 2010, Vol. 44.

26. E.I. Prest, M. Staal, M. Kühlb, M.C.M. van Loosdrecht, J.S. Vrouwenvelder. *Quantitative measurement and visualization of biofilm O2 consumption rates in membrane filtration systems.* Journal of Membrane Science, 2011, Vol. 392.

27. C.P. Koutsou, S.G. Yiantsios, A.J. Karabelas. *Numerical simulation of the flow in a plane channel containing a periodic array of cylindrical turbulence promoters.* Journal of membrane Science, 2004, Vol. 231.

28. Y.L. Li, K.L. Tung. *The effect of curvature of a spacer-filled channel on fluid flow in spiral-wound membrane modules.* Journal of Membrane Science, 2008, Vol. 319.

29. J.S. Marshall. *Discrete-element modeling of particulate aerosol flows.* Journal of Computational Physics, 2008, Vol. 228.

30. L. Shuiqing, J.S. Marshall, G. Liu, Q. Yao. *Adhesive particulate flow: The discrete-element method and its application in energy and environmental engineering.* Progress in Energy and Combustion Science, 2011, Vol. 37.

31. M. Elimelech. *Particle deposition on ideal collectors from dilute flowing suspensions: Mathematical formulation, numerical solution, and simultions*Separation Technology, 1994, Vol. 4.

32. J. A. L. Kemps, S. Bhattacharjee. *Particle Tracking Model for Colloid Transport near Planar Surfaces Covered with Spherical Asperities.* Langmuir, 2009.

33. A. Hernàndez, J. Calvo, P. Pràdanos, F. Tejerina. *Pore size distributions of track-etched membranes: comparison of surface and bulk porosities.* Colloids Surf., 1998.

34. Invitrogen. Latex beads technical overview: http://www.invitrogen.com/

35. K.J. Hwang, H. C. Chen. *Selective deposition of fine particles in constant-flux submerged membrane filtration* Chemical Engineering Journal, 2010, Vol. 157.

36. Hydranautics. Pretreatment and design limits section of help module: RODESIGN, Hydranautics RO System Design Software, VERSION 6.4, 1998.

37. N.A. Fuchs. *The mechanics of aerosols.* New York, Dover Publications, 1989.

38. M. Bikel, I. Punt, R. Lammertink, M. Wessling. *Shrinkage effects during polymer phase separation on microfabricated molds.*: Journal of membrane science, 2010, Vol. 347.

39. R.W. Baker. *Membrane Technology and Applications.* John Wiley and Sons, 2004.

40. C. S. Tsay, A. J. McHugh. *Mass Transfer Modeling of Asymmetric Membrane Formation by Phase Inversion.* Journal of Polymer Science, 1990, Vol. 28 1327.

41. C.A. Smolders, A J. Reuvers, R.M. Boom, I.M. Wienk. *Microstructures in Phase-Inversion Membranes: Part 1. Formation of Macrovoids.* Journal of Membrane Sctence,, 1992, Vol. 73.

42. N. Ochoa, P. Pràdanos, C. Palacio, C. Pagliero, J. Marchese , A. Hernàndez. *Pore size distribution based on AFM imaging and retention of multydisperse polymer solutes. Characterization of polyethersulfone UF membranes with dopes containing different PVP.* Journal of membrane science, 2001, Vol. 187.

43. R.M. Boom, T. Van Den Boomgaard, C.A. Smolders. *Mass transfer and thermodynamics during immersion precipitation for a two-polymer system. Evaluation with PES-PVP-NMP-water.* Journal Membrane Science, 1994, Vol. 231.

44. L. Lafreniere, F. Talbot, T. Matsuura, S. Sourirajan. *Effect of polyvinylpyrrolidone additive on the performance of polyethersulfone ultrafiltration membranes.* Ind. Eng. Chem., 1987.

45.N. Vogrin. *The wet phase separation: the effect of cast solution thickness on the appearance of macrovoids in the membrane forming ternary cellulose acetate/acetone/water system.* Journal of Membrane Science, 2002, Vol. 207.

46. R. Boom. *Membrane formation by immersion precipitation: the role of a polymeric additive.* Enschede , Twente University , 1992.

47. P. Kulkarni, R. Sureshkumar, P. Biswas. *Multiscale simulation of irreversible deposition in presence of double layer interaaction*, Journal of Colloid and Interface Science, 2003, pp. 36–48.

48. J.R. Dunne. *Bacterial adhesion: seen any good biofilms lately?*, Clin. Microbiol. Rev., 2002, Vol. 15.

49. http://www.aquanext-inc.com/en/index.html.

50. T. Miyano, T. Matsuura, S. Sourirajan. *Effect of polyvinylpyrrolidone additive on the poresize and the pore size distribution.* Chem. Eng. Commun, 1993.

51. A.L. Ahmad, K.K. Lau. *Impact of different spacer filaments geometries on 2D unsteady hydrodynamics and concentration polarization in spiral wound membrane channel.* Journal of Membrane Science, 2006, Vol. 286.

52. M. Gimmelshtein, R. Semiat. *Investigation of flow next to membrane wall.* Journal of Membrane Science, 2005, Vol. 264.

53. J. Schwinge, D.E. Wiley, D.F. Fletcher. *Simulation of the flow around spacer filaments between narrow channel walls: Hydrodynamics.* Ind. Eng. Chem., 2002, Vol. 12.

ACKNOWLEDGMENTS

Looking back, it gives me immense pleasure to acknowledge the many people who contributed in various way and by no small measure to the successful completion of this book.

As the old adage goes "A thousand miles' journey, always begins with the first step", my first thought goes to my whole family and my mom, Alfonsina in particular. Thanks for all your unconditional love, and even if we are far away, I feel really blessed every single day, knowing how much you all care.

To my girlfriend Mariapia, words are inadequate to describe how much I thank you for always supporting me and believing in me.

I am also vastly grateful to Francesca. You convinced me to accept the enrollment to Wageningen University. I will never forget your "*E mannala sta lettera!!!*" ... It was the best decision I have ever made.

I am grateful to my first supervisor and promoter of this project, Prof. Rob Lammertink (University of Twente) and the whole *sfi* group. I learnt a lot from you and you remain one big source of inspiration to my development as a scientist. I am also thankful to my co-promoter, Prof. Hans Vrouwenvelder from TU Delft. Thank you for granting me opportunity to undertake this further step in my quest for expert knowledge at the cutting edge of membranes science.

I humbly express my gratitude to my daily supervisor, Prof. Cristian Picioreanu (TU Delft) for his superb guidance, encouragement and support and for critically reviewing this manuscript. Cristian, having worked with you during this time has been just "Amazing" (as you always say)! Your immense passion for everything you do is a powerful engine that leads to discover "new planets" every day. A special thought goes to my other supervisor, (Dr.) Andrea Radu; you are such a nice person and working with you was supercool. Your contribution did not only help to build the numerical model (e.g. computing the particle motion in MATLAB), but it was invaluable for the completion of this manuscript. Thanks again. Finally, to all the people working in the *EBT* group in TU Delft, thanks for the wonderful time spent together.

Surrounded by so many friends and loved ones, it is inevitable that I cannot insert all the names. To all who journeyed with e and supported me in one way or the other, I heartily thank you all!!!